Charlie finds a pattern

in the sandpit

Sarah Reeves

Sarah's Books Pty Ltd, PO Box 1904 Armidale NSW 2350, Australia

First Edition, 2024
First published, 2024

ISBN: 978-1-7635799-0-3
Independently published.

Printed in Sydney, NSW, Australia, if purchased in Australia.
eBook also available.

A catalogue record for this book is available from the National Library of Australia.

This book was created, written and illustrated in Armidale NSW, Australia, which is Anaiwan country.

Links to the Australian Mathematics Curricula

ACARA Mathematics Curriculum link:
Foundation - AC9MFA01 - recognise, copy and continue repeating patterns represented in different ways.

NSW Mathematics K-10 Curriculum link:
Early Stage 1 - MAE-FG-01 recognises, describes and continues repeating patterns.

Victorian Mathematics F - 6 Curriculum
Foundation - VC2MFA01 - follow a short sequence of instructions; recognise, copy, continue and create repeating patterns represented in different ways

Links to curricula:
The Australian Curriculum, Assessment and Reporting Authority (ACARA) Mathematics F-10 Version 9.0
The NSW Education Standards Authority (NESA) Mathematics K-10 Curriculum (February 2024)
The Victorian Curriculum and Assessment Authority Mathematics Foundation to Level 6 V 2.0

This book is dedicated to my Mum.

Thank you for your never ending love and support.

Thank you.

It is a lovely sunny day and

Charlie plays outside

in the sandpit.

Charlie finds a rock.

Then a stick.

Next, Charlie finds

another rock.

And another stick.

And rock.

And stick.

Charlie has found a pattern.

Charlie finds one more stick.

Charlie notices the pattern is

continuing.

Rock, stick, rock, stick,

rock, stick, rock, stick,

rock, stick...

A pattern is something that

reoccurs, repeats

the same order of things

over and over again.

The End.

Other books available in the Charlie series:

Charlie counts to five, on a picnic
ACARA Mathematics Curriculum link AC9MFN02
NSW Mathematics K-10 Curriculum link MAE-RWN-001 & MAE-RWN-002
Victorian Mathematics F - 6 Curriculum link VC2MFN02

Charlie explores whole number 7
ACARA Mathematics Curriculum link AC9MFN06
NSW Mathematics K-10 Curriculum link MAE-RWN-01
Victorian Mathematics F - 6 Curriculum link VC2MFN06

Charlie loves to share, exploring odd and even numbers
ACARA Mathematics Curriculum link AC9MFN06
NSW Mathematics K-10 Curriculum link MAE-FG-02 & MA1-FG01
Victorian Mathematics F - 6 Curriculum link VC2MFN06 & VC2M3N01

Charlie looks into greater than, less than and equal to
ACARA Mathematics Curriculum link AC9MFN06
NSW Mathematics K-10 Curriculum link MAE-RWN-02
Victorian Mathematics F - 6 Curriculum link VC2MFN01

Charlie goes for a walk, introducing algebra, with oranges
ACARA Mathematics Curriculum link AC9MFA01 & AC9M1A02
NSW Mathematics K-10 Curriculum link MAE-FG-02
Victorian Mathematics F - 6 Curriculum link VC2MFA01 & VC2MFA02

Other books available in the Charlie series:

Charlie loves to cook, exploring measurement
ACARA Mathematics Curriculum link AC9MFM01
NSW Mathematics K-10 Curriculum link MAE-CSQ01
Victorian Mathematics F - 6 Curriculum link VC2MFM01

Charlie explores fractions
ACARA Mathematics Curriculum link AC9M2M02
NSW Mathematics K-10 Curriculum link MAE-GM-03
Victorian Mathematics F - 6 Curriculum link VC2M2M02

Charlie finds Geometry, in nature
ACARA Mathematics Curriculum link AC9MFSP01
NSW Mathematics K-10 Curriculum link MAE-2DS-01 & MA1-2DS-01
Victorian Mathematics F - 6 Curriculum link VC2MFSP01

Links to curricula:
The Australian Curriculum, Assessment and Reporting Authority (ACARA) Mathematics F-10 Version 9.0
The NSW Education Standards Authority (NESA) Mathematics K-10 Curriculum (February 2024)
The Victorian Curriculum and Assessment Authority Mathematics Foundation to Level 6 V 2.0

Made in the USA
Monee, IL
05 November 2026

67380370R00019